Ground Radar Systems of the German Luftwaffe to 1945

FuMG 39 T along the Atlantic seawall in southern France. A 20 mm Flak 38 can be seen in place in the background, its purpose to protect the site against strafing attacks.

Werner Müller

Schiffer Military/Aviation History
Atglen, PA

Sources

Hoffmann-Heyden, Adolf-Eckard: "Die Funkmeßgeräte der deutschen Flakartillerie 1938-1945". Volume 3 of "Bücherei der Funkortung", Verkehrs- und Wirtschafts-Verlag, Dortmund

Koch, Horst-Adalbert: "Flak", Podzun-Pallas Verlag, Friedberg 3, 2nd ed. 1965

Lusar, Rudolf: "Die deutschen Waffen und Geheimwaffen des Zweiten Weltkrieges und ihre Weiterentwicklung". Lehmanns-Verlag, Munich 1971

Niehaus, Werner: "Die Radarschlacht 1939-1945". Motorbuch-Verlag, Stuttgart 1977

Renz, Otto-Wilhelm v.: "Deutsche Flugabwehr im 20. Jahrhundert", Mittler-Verlag, Frankfurt 1960

Reuter, Frank: "Funkmeß", "Die Entwicklung und der Einsatz des RADAR-Verfahrens in Deutschland bis zum Ende des Zweiten Weltkrieges", Westdeutscher Verlag, Opladen 1971

Trenkle, Fritz: "Die deutschen Funkmeßverfahren bis 1945", publication of the Dr. A. Hüthig Verlag, Heidelberg 1986

L.Dv., 400/8 Supplement: "Richtlinien für die Ausbildung an den Funkmeßgeräten (Flak), February 1944

Photo Credits

Bundes-Archiv Koblenz (26), Fritz Trenkle (46), AEG (4), Museum für hist. Wehrtechnik, Nuremberg (1). From the private collections of: Eberhardt Reimer (2), Hermann Reinecke (2), Dr. Hans-Dietrich Nicolaisen (1), Alfred Otte (2), Michael Schmeelke (5).

A word Beforehand

Naturally, such a comprehensive theme as German radar systems cannot be covered in detail within the framework of this book. Our primary goal is to present the interested reader with an overview of those ground-based radar systems operated by the Luftwaffe up to 1945. Accordingly, the main focus is on equipment operated by the Flak for air defense, on fighter direction and air control systems and on equipment used for protecting civilian centers.

Those who are interested in specific details of these systems are encouraged to turn to the material and bibliography in Fritz Trenkle's book "Die Deutschen Funkmeßverfahren bis 1945". Marketing & Technik-Verlag, 7707 Engen 5. The author has provided us with many of the pictures from his collection for publication in this book, for which we would like to thank him at this point. Another good source is Volume 3 of "Bücherei der Funkortung", titled "Die Funkmeßgeräte der deutschen Flakartillerie (1938-1945)" by A. E. Hoffmann-Heyden and published by State Secretary Prof. Leo Brand at the Verkehrs- und Wirtschafts-Verlag in Dortmund. Here the technically-minded reader will find additional details regarding the construction and electrical layout of the systems, assuming that a local bookshop can use its extended network to find a book that has been out of print for nearly 40 years.

An FuMB 27 radar observation site with traversing "Chinese base", on the island of Heligoland after the air raid on 18 April 1945.

COVER PICTURE:
The cover picture shows an FuMG 39 T(D2).

Translated from the German by Don Cox

Printed in China.
ISBN: 0-7643-0567-0

This book was originally published under the title,
Waffen Arsenal-Bodenfunkmessgeräte der Deutschen Luftwaffe bis 1945
by Podzun-Pallas Verlag.

We are interested in hearing from authors with book ideas on related topics.

Published by Schiffer Publishing Ltd.
4880 Lower Valley Road
Atglen, PA 19310
Phone: (610) 593-1777
FAX: (610) 593-2002
E-mail: Schifferbk@aol.com
Please write for a free catalog.
This book may be purchased from the publisher.
Please include $3.95 postage.
Try your bookstore first.

Origins of Radar Technology

The origins of radio ranging technology, more commonly known as radar technology, had their beginnings back in the year 1900. (The word "radar" is an acronym from the English designation "radio detection and ranging", or using radio waves to acquire and determine the distance of objects.) At the time, a student in Bremen by the name of Christian Hülsmeyer came across the discoveries of the physicist Heinrich Hertz, who had taught at Karlsruhe University and in 1887 proved through experimentation there that a field of electromagnetic waves was generated around a radio inductor when it was used as a transmitter. These waves were reflected from metallic surfaces just as rays of light were reflected by a mirror. Hülsmeyer attempted to exploit this echo effect of electric waves by developing a system for ships which would warn of collisions in bad weather. Using a homemade radar device set up on the Rhine River near Düsseldorf, he demonstrated how electric waves were reflected off the sides of ships. Unfortunately, however, the time was not yet ripe for such an invention; the shippers were still dependent upon Marconi and his monopoly and did not recognize the difference between radio communications and radio detection and ranging.

Even the radar experiments of Scherl and Hans Dominik, a well-known author of science-fiction novels, found no further support during the World War One years. It wasn't until 1934 that Dr. Kühnhold founded the GEMA, from "Gesellschaft für **E**lektroakustische und **M**echanische **A**pparate" - Association for Electroacoustic and Mechanical Equipment (not to be confused with today's Gema organization). It was there that for the first time work became focused on developing a radar system. Efforts were primarily concentrated on using an operating frequency in the VHF range (125 MHz), since the required directional antennas could be mounted on a rotating base together with the other necessary equipment. On 24 October 1934 naval officers and civilian officials were given a demonstration in Lübeck Bay, whereby rays were reflected back from the test boat "Welle", at rest some 12 kilometers away. Even a Junkers W34 which happened to be flying by at around 700 meters' altitude reflected the radiated impulses back to the receiver, which was located some 200 meters away from the transmitter.

Although the Marine(Navy) was impressed by the demonstration and approved 70,000 RM for further development, the military displayed little initial interest in radar. Nevertheless, work continued in this area at GEMA and, in the meantime, also at the Lorenz and Telefunken companies. These efforts led to a system being built at GEMA, which with its "Tannenbaum" antenna functioned with a wavelength of 2.4 m (125 MHz) and which could effectively be called the original prototype of the subsequent "Freya" systems. With this device, it was possible to acquire aircraft at a distance of 40 to 75 kilometers, depending on altitude.

At the end of 1938 the first of the improved A1 systems, called Freya, was delivered to the field units. The second system went into operation as German troops marched into Czechoslovakia, during which a great deal of knowledge was gained with regard to the effects of terrain on the system. At the beginning of the Second World War on 1 September 1939 there were a total of eight Freya systems in operation, stationed on a handful of islands in the North Sea. Anti-aircraft targeting radars, or FuMG (from Funkmeßgerät, or radar) were not yet in service when the war broke out. The absence of such systems was to be sorely felt within the first few weeks of the war.

Right:
One of the first experimental radar systems by the DVG company near Munich was a continuous wave radar, i.e. transmission and reception took place simultaneously, whereby the transmitting and receiving antennas had to be kept well separated from each other. The system worked on a 10 cm wavelength and when targeted against a steamer on the Ammersee, demonstrated a range of one kilometer. By using a Doppler effect it was also possible to determine the speed of a target.

The Anti-Aircraft Artillery's Need for Radars

The firing accuracy of the heavy anti-aircraft artillery's (AAA) guns was only possible by making precise calculation of the lead required for hitting air targets as they moved through their three-dimensional space. Using specialized calculating equipment - fire control systems - the distance, azimuth and altitude of the target was initially determined using optical means. Due to the need for constantly updated target location information, these systems computed the necessary lead and angle data for the guns by determining the angular velocity (with the Kdo.Hi.Ger. 35) or by geometric-linear methods (such as with the Kdo.Ger. 36 and the virtually automatic Kdo.Ger. 40). These systems also had to take into account such things as load delay time (generally around 3 seconds), the height of the gun above sea level, the weight class of the round being fired, the powder temperature and dampness, trajectory deviation, plus weather conditions like air density, wind speed and direction. The data resulted in alterations to the fuse burn time, the firing azimuth and the firing altitude, i.e. the overall elevation of the barrel. (for further information see "Sound Locators, Fire Control Systems and Searchlights of the Heavy Flak" in this series).

An optical acquisition of targets at night could only be accomplished using searchlights, which in turn were directed using sound locating cones. What happened, then, when the day or night skies were covered in dense clouds, or no searchlights were available on a particular night, or they couldn't locate the target? In these cases the heavy gun units were forced to rely on their own audio locating dishes; these could only determine the direction of the target, not the distance. If two of these acoustic location systems were available at two separate locales, the approximate distance to the target (to within roughly 10 km) could be determined by the process of triangulation, but even then the accuracy was affected by weather conditions. However, this method of target location required a considerable amount of preparation time for both the equipment and the crews, so that targeting the ever-faster and higher-flying approaching aircraft eventually became impractical using sound location. The ammunition-hungry barrage fire method also met with little success; this involved throwing up a curtain of fire at an estimated altitude and distance. These deficiencies became particularly apparent in homeland defense operations, after the British, following a daylight raid on Wilhelmshaven in which they suffered crippling losses, resorted to penetrating German airspace only at night. On 18 December 1939 three of the RAF's Wellington bomber units were picked up by the Freya radars set up on the islands of Heligoland and Wangerooge. The information was passed on to a Jagdgeschwader in Jever. According to a report by the Oberkommando der Wehrmacht (OKW), the British lost 34 aircraft downed to fighters as a result of the early warning radar report. For the first time, the Freya early warning system had demonstrated its worth.

The AAA units, however, were in desperate need of targeting systems which could operate independent of visual acquisition and in all types of weather. The non-optical data obtained by this system needed to be processed by the standard Folgezeigerübertragungsgerät 37, which was at the time a highly refined follow-the-pointer target tracking data transmission system. In addition, there was a need for accurate location fixing out to at least 25 km with an azimuth accuracy of ±1/- and vertical accuracy of ±1/16°. The system should be as immune as possible to enemy jamming measures and have an identification system for differentiating between friendly and hostile targets. Not only that, but it also needed to be easily road transportable, weatherproof, unaffected by temperature and shockproof. Electricity was to be provided by the standardized grid system of 220 V 50 Hz or supplied by a transportable generator providing 380 V 50 Hz. These requirements, however, far exceeded the state of development in 1939/40.

The first AAA targeting radar produced by the firm of Lorenz was the A2-Gerät FuMG (Flak) 38 L Kurfürst. Using the cross-base of an 88 mm anti-aircraft gun, the barrel was replaced by a rotatable tubing mast to which two sliding parabolic dish antennas were affixed one above the other and each having a diameter of 2.4 meters.

FuMG 38 L Kurfürst

The A2-Gerät Kurfürst made by the C. Lorenz company, with its rotating upper section, was fitted to the cross-base of the 88 mm Flak gun and as such could be towed on the Sonderanhänger 202 trailer. A hollow tube mast was fitted in place of the gun's barrel onto which two parabolic dishes were attached, one for sending and one for receiving the radar signal and each having a diameter of 2.4 meters. A slide rail permitted these dishes to be adjusted up or down, and they could also be tilted by means of a hinge mechanism. The dishes were covered in wire mesh to keep the wind resistance to a minimum when the mast was extended to its full height of 6.5 meters. The system operated on a wavelength of 62.4 cm and had a detection range against airborne targets of 8 - 12 km; its range accuracy was ±100 meters and the azimuth/height accuracy was ±3-4°.

The FuMG Radar Systems of the AAA

The previously mentioned Freya system built by the GEMA company and designated "A1 Gerät" was an early warning radar, but could not function as a target acquisition radar. The companies of C. Lorenz and Telefunken developed such a system, with the former's developmental type designated as A2 and the latter's as A3.

A note on the abbreviated designations for the systems: beginning in 1942 series production developments also carried their official designations in addition to their codewords, e.g. FuMG 39 L = Funkmeßgerät (radar), introduced in 1939, manufactured by the Lorenz company (G=GEMA, T=Telefunken, S=Siemens, A=AEG). In the Luftwaffe the numbers 62-99 and design classifications of A, B, C and D replaced the year designator, e.g. FuMG 62A Würzburg.

FuMG 39 L Kurpfalz

The FuMG 39 L Kurpfalz was a follow-on development of the FuMG 38 L Kurfürst with a higher transmission output and improved ranging accuracy. It was initially introduced as a production batch of 20 systems and supplied to AAA units in the Ruhr district and along the Channel coast for front-line evaluation.

The antenna, consisting of two wire mesh parabolic dishes in a stacked configuration and each having a diameter of 2.4 meters, could be rotated and tilted on top of the operations van to which they were attached. The fork-shaped antenna support and the two dishes could be folded down to the roof of the operations van and secured. Transportable on a two axle trailer, the system's range was somewhere between 10 - 25 km, ranging accuracy was ±40-50 km, accuracy was ±2-3° for azimuth and vertical accuracy was ±3-4°.

Above: When deployed, the operations van of the FuMG 39 L Kurpfalz system was stabilized by side-folding support arms. The antenna constituted two 2.4 m diameter parabolic dishes arranged one above the other.

Right: The parabolic dishes could be folded down and secured onto the roof of the operations van using a hinge mechanism. They were raised upright with the aid of a hydraulic pump.

FuMG 40 L Kurmark

The FuMG 40 L Kurmark was initially planned only as an interim solution, further development on which would be stopped for the higher priority FuMG 41 L; this effectively meant that the Kurmark was the last radar system built by Lorenz for deployment with AAA units. Externally it differed little from the FuMG 39 L. Improvements were achieved by increasing the transmission power to 50 kW and thereby increasing its detection range to 25 - 40 km, in search mode as far as 50 km. The range accuracy was ±30-40 m and its azimuth/vertical accuracy was ±10-12/16°.

The Telefunken company also became involved in the development of radar systems for anti-aircraft units beginning in 1936. The first A3 system, Darmstadt, was a test prototype and not designed to be mobile. However, the Darmstadt can be considered the immediate predecessor of the subsequent and much improved Mainz and Mannheim systems, discussed in detail later. The system differed considerable from the A1 and A2 systems both in its technical construction (which cannot be covered fully here) and purely from external appearances. It had no "mattress" or "flycatcher" antenna, only a single dish of 3 meters diameter and which operated in simultaneous mode, i.e. the antenna surface was utilized by both the transmitter and the receiver. The system had a range of 8 - 10 km, a range accuracy of ±100 m and an azimuth/vertical accuracy of ±1/4°, giving it much more precision than the A2 system. The A3 system was also equipped with remote controls which permitted an operations center to acquire and track targets at a location separate from the dish. Curiously, this feature was not incorporated into successive systems, not becoming popular until much later.

The FuMG 40 L Kurmark, an improved version developed by the Lorenz company, different little in appearance from its predecessor types. Notice the receiver located just behind the upper dish.

FuMG (FLAK) 39 TA, -C, -D Würzburg (FuMG 62 A-D)

Beginning in 1937 Telefunken also became involved in the development of a small radar system for short-range location and identification. This was the FuMG 39 T Würzburg A. In the fall of 1939 the system was introduced with the designation of FuMG 62 Würzburg. It rested on a two-axle chassis, which was stabilized during operations by four lifting jacks. The parabolic dish, having a diameter of 3 meters, was mounted together with its tip rocker to the system housing. The dish could be tilted from 0° to 90° and rotated through a full 360°. The radar had a range of up to 40 km with an accuracy of ±80-120 m. The azimuth/height accuracy was ±1.5-2°. Yet these capabilities did not sufficiently provide adequate data to the heavy AAA batteries' command and control systems. Nevertheless, with the aid of a data transfer system (known as Malsi) several batteries were linked up to the relatively few FuMG systems which were initially available, despite having entered full-scale production. This enabled these batteries to throw up a relatively concentrated, directed fire against approaching enemy aircraft.

The system was either modified or built from the outset as the FuMG 39 T/C Würzburg C with a precision send-and-receive signal kit (the C-Zusatz) developed by Telefunken in 1940. In place of a fixed dipole the system was now fitted with a running dipole, a "turnstile", which gave it an azimuth accuracy of ±0.45° and a vertical accuracy of ±7-9/16°. With the introduction of the EAG 62 Emil "D-Zusatz" distance measuring accuracy was improved to ±25-40 m and at the same time provided a direct feed of ranging data to the ranging processor of the Übertragungsgerät 37, from whence the data would be sent on to a command and control system at a AAA battery. The system, known as the FuMG 39 T/D Würzburg D (later FuMG 62 D), began arriving at the front in 1942. By the war's end approximately 4,000 FuMG 62 systems had been manufactured, making it the standard radar system with the AAA forces.

The above photo shows one of the first G 38 39 T/A Würzburg A systems (later FuMG 62 A). The fixed dipole protrudes from the center of the dish, while on either side of this can be seen auxiliary dipoles for IFF reception. The IFF system necessitated aircraft being fitted with the FuG 25 Zwilling.

Below: The wooden traverse plate of the FuMG 62 Würzburg 39 T D2 was located 35 to 45 cm above ground, depending on the levelling height needed. In travel mode, both sides of the traverse plate could be folded up, where they were held in place by four braces. When in operating mode, eight extendable wooden beams stabilized the traverse plate. The "sidecar", an extension protruding from the right of the rotating system housing, contained the primary components and operating controls with a control seat, complete with heated leg shrouds for the B2 operator. The upper portion of the 3-meter diameter antenna dish could be tipped 180° to the rear for transport on the two-axle Sonderanhänger 104 trailer.

This is an FuMG 39 T D2 or D3 with its wooden traverse plate. The crew, seen in position, are as follows (from left to right): The B6 seated at the elevation quadrant passed the altitude information by voice radio to a command and control center or to an associated searchlight unit. Next to him, the B5 sitting at the bearing indicator instrumentation passed this information on by radio as needed. The B1 stands at the rangefinder, a cathode ray tube(CRT). Next to him is the B3 standing at the height tracking scope, while the B2 sits in the control seat, where he controls the elevation angle and bearing of the system. If the Übertragungsgerät 37 were not operating, the B4 would stand next to the B1 at the rangefinder to pass on ranging information via radio.

Above: This FuMG 39 T was assigned to 2 Batterie of I/42 (mot.S.), where it is seen here set up at an exposed site in October of 1941 near Stettin. The 88 mm guns of this particular AAA unit were the only ones in the entire Wehrmacht to be mounted as self-propelled guns on VOMAG chassis.

Below: This is probably the site of a prototype system. The dish is the same style as the small Würzburg A-,B-, C- and D-systems. Based on scale comparisons with the wooden support wall (which has a height of approx. 1.90 m - 2.30 m) and the vehicle, it appears that the diameter of the antenna dish would be larger than 3 meters, possibly the 4.5 meter type as on the FuMG 68 Ansbach. The light colored speck on the upper edge of the dish is an anchor point for a larger IFF antenna. In the background to the left is a 150 cm searchlight.

Above: This FuMG 39 T D has been set up here along France's Atlantic coast in a well-fortified 88 mm gun emplacement.

The two individual single-axle components of the Sonderanhänger 104 have been pulled away from this FuMG 39 T D4. The axle components were interchangeable and could be used both fore and aft. D4 systems were built with a metal traverse base which, as can be seen in this photo, were stabilized with metal supports. A cross antenna protrudes from the covered dipole, the purpose of which was to switch frequencies in case of jamming interference.

The Halifax, Lancaster and Wellington bomber markings painted inside the reflector dish of this Würzburg system indicate successful kills in which this radar crew played a part. The dipole is attached to the conical dipole shaft by means of a screw cap. Two leads run out from the dipole in the center of the dish. One serves as the power lead for the drive motor and was plugged into the power box with a three-pronged plug, while the other was attached to the control system by a six-pronged plug.
A handful of individual components of the FuMG 62 A-D Würzburg was assigned covernames: +8.3 kV power unit = "Dachs", -2.3 kW + 350 V power unit = "Katze", -50 V power unit = "Hund", transmitter = "Lokomotive", Sü 62 component = "Eidechse"", ZFV 62 = "Zobel".

The Würzburg C system had a rotating antenna and a new ANG 62 scope. Located on the inner lip of the reflector dish is one of the IFF dipoles with its protective cap on. Just opposite this, on the dish's outer lip, is one of the two snap locks which hold the two dish halves together. These are two of the crew members; the B6 sits in front of the elevation quadrant and the B5 at the azimuth scale. Both were to pass their respective data on to designated command and control centers or searchlight units by voice radio in the event that the Übertragungsgerät 37 was not operating.

Protected by a canvas cover, the B1 is at work on the CRT of the precision ranging system, the B2 is at the horizontal tracking scope and the B3 at the vertical tracking scope.

Below: On the right and left edges of this Würzburg D reflector dish can be seen the connections and cap mounts for the IFF system, using the "Kuh" transmitter and "Gemse" receiver. The associated antenna is attached to the antenna cap on the dipole.

Above left: Here we see the B2 in the foreground, sitting in the control seat where he's moving the system on its lateral axis. He is looking into the range and bearing indicator in front of him. Behind him stands the B3 at the elevation monitoring scope and is operating the elevation crank on the radar housing.

Above right: Again the B2 is seated in the control seat, the so-called "side-car", which is fitted with a backrest. The leg shrouds contained foot warmers and foot switches for switching the antenna on and off. The monitors are covered with weather protectors.

Left: Here the B2 is wearing earphones, indicating that the anti-jamming method "Nürnberg" is being used to counter "Düppel" (chaff), i.e. an audio enhancer is being used to feed the Doppler modulation of the propeller sounds acoustically to the B2's headphones. Volume and clarity of the sound indicated on the CRT had to be optimal in order to achieve a good fix.

This Würzburg D system of a railway AAA unit is anchored to a special rail car of the Reichsbahn. The system platform has been widened, braced on the sides, and is fitted with railing. The EAG 62 distance measuring device is attached to the extension on the right next to the radar housing. This device was used on the FuMG 62 D from 1942 on. The EAG 62 "Emil" D-Zusatz provided improvements in ranging accuracy, a constant zero calibration, simplified operations and also enabled the data to be fed directly to the ranging processor of the Übertragungsgerät 37. In addition to the radar, the other half of the rail car holds the crew's quarters.

FuMG 65 Würzburg-Riese

The FuMG 65 Würzburg-Riese (Würzburg Giant) was developed from the FuMG 62 D and was given the latter's electrical ranging components. However, in order to increase the range to 50 -70 km, the system was fitted with a dish having a diameter of 7.5 meters, also known as "Nashorn" (Rhinoceros). This enabled aircraft to be tracked out to 40 - 60 km, with raids being acquired as far out as 80 km. Range accuracy was 15 - 20 meters. With an azimuth and vertical accuracy of around ±0.2° and ±0.1°, respectively, the system was not only suited for the signals communication troops and fighter control, but also found a role with both the Flak and the Marine. It was sited on flak towers, in large batteries and on ships. Mobile systems, designated as FuMG 65 E Würzburg-Riese E, were installed in limited numbers on rail cars, although these had a somewhat modified dish drive. The entire system weighed around 18 tons. The moveable portion, weighing 11 tons and consisting of the dish and its tip rocker plus the operations van, was operated with the assistance of an AEG Leonard control system. The dish and rocker on the one side and the operations van on the other side of the cross-member support virtually balanced each other out, so that no undue stress was placed on the support itself. The tip rocker bearings on either side of the dish were attached at the dish's center of gravity, making vertical movement of the dish considerably easier. The dish itself was made of metal alloys which were in common use from the construction of airships in Friedrichshafen. Despite the equal distribution of weight it was not possible to operate the drive unit directly by hand. Instead, the system was controlled with the assistance of the AEG remote operating unit, which cannot be described in detail here. Basically, hand wheels were used to operate the drive motors, which permitted the dish to be moved at any desired rate of speed. This remote operation feature permitted the pivoting base and its drive motors to be located separately from the operations center, meaning that the latter could be housed in a bunker, for example, such as when sited on flak towers. This control system was also used later with the FuMG 68 and FuMG 76 (discussed later in this book).

Since the concentrated focus of radar waves rarely made it possible to acquire rapidly moving targets using the FuMG 65 without data from a Freya system, some "Giants" were fitted with a Freya supplemental search kit of 2.4 m wavelength, located on either side of the "turnstile" (see photo below).

The FuMG 65 Würzburg-Riese rested on a pivot shaft, which in turn generally had a concrete base. The operations van and the tip mechanism and dish opposite the van, which virtually balanced each other out, rotated around this pivot shaft.

The operations van of a FuMG 65 Würzburg-Riese; the tip bearing and the external parts to the elevation drive can also be seen to good effect. This system is located at the former "Diogenes" Luftwaffe control center at Schaarsbergen west of Arnhem. It can be seen there today, although a few of the critical components (e.g. the dipole and the radar equipment housing in the operations van) are missing.

Above: The reflector dish had a diameter of 7.5 meters. For an antenna, the same DAS 62 dipole was used as found on the FuMG 62 D. It stuck out a distance of 1.72 meters from the reflector's surface. The reflector was supported by two rocker bearings located on either side through the dish's center of gravity. This meant that the radar's elevation drive motor was only required to supply enough power to tilt the dish.

Below: This photo shows the size of the Würzburg Riese's reflector to good effect. The absent dipole on the previously shown system at Schaarsbergen is indicated by a wire pyramid.

Two photos of a FuMO Würzburg-Riese, which is the naval designation for the FuMG 65. The radar system in these photos belonged to 3/III FMA NANTES, near Saint Servais above La Baule, and was used for coastal observation. Photo taken in September 1943.

Above: A bombed out radar of 3/M.Fl.A. 242 on the island of Heligoland following the raid of 18 April 1945. This probably was a Würzburg-Riese, which along with its entire bunker took a direct hit.

Below: This is a Würzburg-Riese reflector dish on a new, mobile two-axis pivoting base. The radar with its 7.5 m Nashorn reflector was part of the "Hawaii II" multi-track direction system, an A4 remote guidance system for the "V" missiles.

A Würzburg-Riese and a FuMG 39 T Würzburg were set up on the control tower of the Turmflakabteilung 123 in Berlin's Tiergarten. Light anti-aircraft guns were positioned at the corners of the lower platform to provide protection against strafing attacks.

On the left is a Würzburg-Riese E, FuMG 65 E, from one of the railway air raid warning companies set up in 1942. It is fixed to a rail car, which during operations was stabilized on its sides.

Center: The train of a railway air raid warning company seen from the air.

Below: A Würzburg-Riese broken down and being transported on rail cars to the Eastern Front for control of night fighter operations. (We ask that you pardon us for the reproduction quality of these photos, but feel they should be included because of their unique subjects.)

FuMG (FLAK) 40 T MAINZ (FuMG 63)

The FuMG 63 Mainz was produced in the years 1940/41 in a series of 51 units, initially as an interim type before the follow-on FuMG 64. The operations van was the same as on the Lorenz FuMG 39 L and 40 L systems. The 3-meter dish with its revolving dipole, the "turnstile" and its pivot mechanism was attached to the roof. Despite a high bearing accuracy several defects cropped up, so that this radar system never really proved superior to the Würzburg systems. The range of the FuMG 63 was around 25 - 35 km depending on aircraft type and altitude of the target. Accuracy values for distance were ±10-22 meters, for azimuth at ±0.1° and for height at ±0.3-0.5°. Production radar systems were successfully used from 1942 on in the Halle-Leuna protection region.

Above: The FuMG 63 Mainz 40 T made use of the same operations van as with the FuMG 39 L Kurpfalz and FuMG 40 L Kurmark.

Below: Here the operations van of a FuMG 63 Mainz has been entrenched and hidden behind an earthen wall. The IFF antenna can be seen on the 3-meter dish.

FuMG (FLAK) 41 T Mannheim (FuMG 64)

The FuMG 41 T Mannheim, later designated as the FuMG 64, was tested for the first time in the summer of 1942 and at the time was probably the most advanced system available for the AAA units and remote missile guidance. In comparison with the Würzburg system, however, it required a significantly higher degree of technical application. For example, it needed 150 tubes alone, three times more than installed in the Würzburg system. Direction finding was accomplished by means of instruments in place of the CRT. Instead of a fixed operations van the system stood on a frame of the traverse base about 3 meters wide; the base was in turn attached by means of its pivot shaft to a cross platform. On the frame's front side were located the non-serviceable power supply units. To the rear were the operating components such as directional controls, gauges and staggered fuses arranged along a faceplate. The crew members sat on a bench in front of this. They were covered on all sides by a protective shroud. The swivel hinge and drive for vertically slewing the 3 meter parabolic dish were attached to the upper crossbeam of the frame. When transporting the system on the two-axle Sonderanhänger (Sd.Ah.) 204 the upper half of the reflector was uncoupled and folded back down under the cover.

The first A-series had a stand on the right hand side of the radar system which held a supplemental system called the "aquarium". Here a point of light was projected onto a transparent panoramic map showing the location of the target with map distance and azimuth to the target. This stand was dropped on the B-series from 1944 on, since it had been made superfluous by improved optical and electrical plotting options. Now the cabin contained a mechanical horizontal range and elevation computer, which was operated by the distance measuring and elevation drives. Through this the data for the horizontal range and height could be fed to the command and control systems or data conversion systems with the aid of the Übertragungsgerät 37 or passed off via radio. Distance accuracy of the Mannheim system was ±10-15 meters at a range of 25 - 35 km, with the azimuth and vertical accuracy at ±0.15°.

Due to the high triangulation precision reflections from the ground led to errors in elevation angles which were not noticed in earlier, less precise systems. These errors were eliminated by a so-called "bottom lip". This was a narrow-mesh wire net, 2.5 meters long, which was attached to the lower edge of the antenna reflector. However, if the FuMG 64 was sited in surrounding earthworks having a diameter of around 8 meters and a rampart height of 1.8 to 1.9 meters, the earthworks had the same effect as the "bottom lip" and at the same time offered the crew protection against shrapnel.

Below: Two FuMG 64 Mannheim 41 T A-series with the control stand on the right side, photographed at an airfield in Denmark. In place of a fixed operations van with moving antenna, this system combined both parts in a rotating box which was set on a pivot shaft onto a cross platform.

This shows another view of the FuMG 64 Mannheim, FuMG 41 T, which was developed in 1941 by Telefunken for the AAA and produced in series from 1942 on.

Above: This Mannheim crew is carefully working on enriching their cooking pot by planting vegetables around their radar site. On the right side of and linked to the FuMG 64's reflector is an sighting collimator. The radar controller could use this sight to direct the crew to a target which he'd acquired optically. He also had his own scope available to him for monitoring the crew, which showed ranging and direction finding data.

Below: The two FuMG 64 Mannheim and 62 Würzburg systems here are sited in the well constructed facilities of a 88 mm heavy battery, of which a few guns can be seen in the background. The earthworks around the FuMG had a diameter of around 8 meters and a height of 1.8 to 1.9 meters. This prevented interference from height finding errors and the system was able to function without the so-called "bottom lip" (see next page). A Flakrichtgerät 40 A anti-aircraft control system with its two collimators can be seen next to the radar sites.

Above: Perceptible height finding errors cropped up for the first time due to the high DF precision of the FuMG 64, caused by waves bouncing back from off the ground. In order to shield against these interfering reflections a so-called "bottom lip" was attached to the lower rim of the antenna dish. This was an existing part of a cylinder jacket of around 2.5 m length made of narrow-mesh wire.

Below: Yet another photo of a FuMG 64 Mannheim with its "bottom lip", which was fitted when the system was set up without earthworks. The radar controller's stand can be clearly seen on the radar unit's right side, although this was only found on the A-series. This "aquarium", as it was called, could be dropped because of optical and electrical plotting options developed later.

FuMG 68 Ansbach

Heavy batteries operating with six or more AAA guns revealed the need for a radar system with a larger reflective dish than just 3 meters' diameter to improve range and DF accuracy. Such as system, however, needed to be mobile, meaning that the dimensions of the dish had to be somewhere between those of the FuMG 62 D and the FuMG 65 Würzburg-Riese. The electrical plotting components were to be separate from the reflector dish in order to better protect the crew in bunkers, safe from bomb fragments from the more and more frequent appearance of ground attack fighter-bombers. These requirements resulted in the evaluation in 1944 of the FuMG 68 Ansbach, in development since 1942 and having a 4.5 meter parabolic reflector mounted on a pivot base. The unit was transportable using a two-axle Sd.Ah. 204 trailer. During transport two segments of the reflector dish could be uncoupled and folded together toward the dish center. The pivot column was folded over horizontally by a stem guide with two toggle joints. In its operating state, the reflector was operated via an AEG-Leonard remote control system from the Bedienungswagen 68 Bayern operations van either housed in a bunker or entrenched up to 30 meters away. The van contained the operating equipment. The range against individual targets was around 25 - 35 km, with search range against aircraft formations increasing to 60 -65 km. Distance accuracy was ±30-40 meters and the triangulation accuracy was around ±0.2°. The unit operated in the 335 - 430 MHz frequency range on the 53.6 cm wavelength. Only a small series of this system was produced, but these never saw operational service.

Two photos of the FuMG 68 Ansbach. It consisted of a Drehstand 68 pivot stand with the remotely controlled 4.5 m reflector dish and the Bedienungswagen 68 operations van (below). The perforated parabolic dish rested with its pivot shaft on a cross base.

The picture on the right shows the Bedienungswagen 68 detached from the FuMG 68 Ansbach. It could operate up to 30 meters away from the reflector dish and contained the crew and operating equipment, protected against explosion fragments. From here the motors necessary for controlling the movement of the dish were driven with the aid of the AEG-Leonard remote control system. In addition, the azimuth and elevation indicator from the Übertragungsgerät 37 on the pivot stand was linked to the operations van by a cable.

In order to be transported on the Sd.Ah. 204 two segments of the overly wide reflector had to be detached and folded in toward the center of the dish. The pivot shaft was lowered to its horizontal transport position and raised to its operating position by hand. See photos.

Two events which had a major impact on the development of German radar technology should be mentioned at this point.

The first was an occurrence during the night of 27/28 February 1942, when in a surprise raid near Bruneval on the Cap d'Antifer near Le Havre the British captured the most critical components of a radar system hitherto unknown to them. It was a Würzburg 62 A system, which through aerial photographs they'd discovered positioned next to a Freya station already known to them. Because of the captured pieces (dipole, transmitter-heterodyne oscillator, intermediate frequency component) it was now possible to develop jamming countermeasures against these systems. Chief among these were the so-called "chaff strips", called "Window" by the British. These were strips of tinfoil which, when cut to one-half the length of the radar's wavelength, reflected back the radar signals as an aircraft would. Since the British now knew that the German Würzburg systems operated on 560 MHz (thanks to the captured parts), the first "Window" operations followed in short order. During the

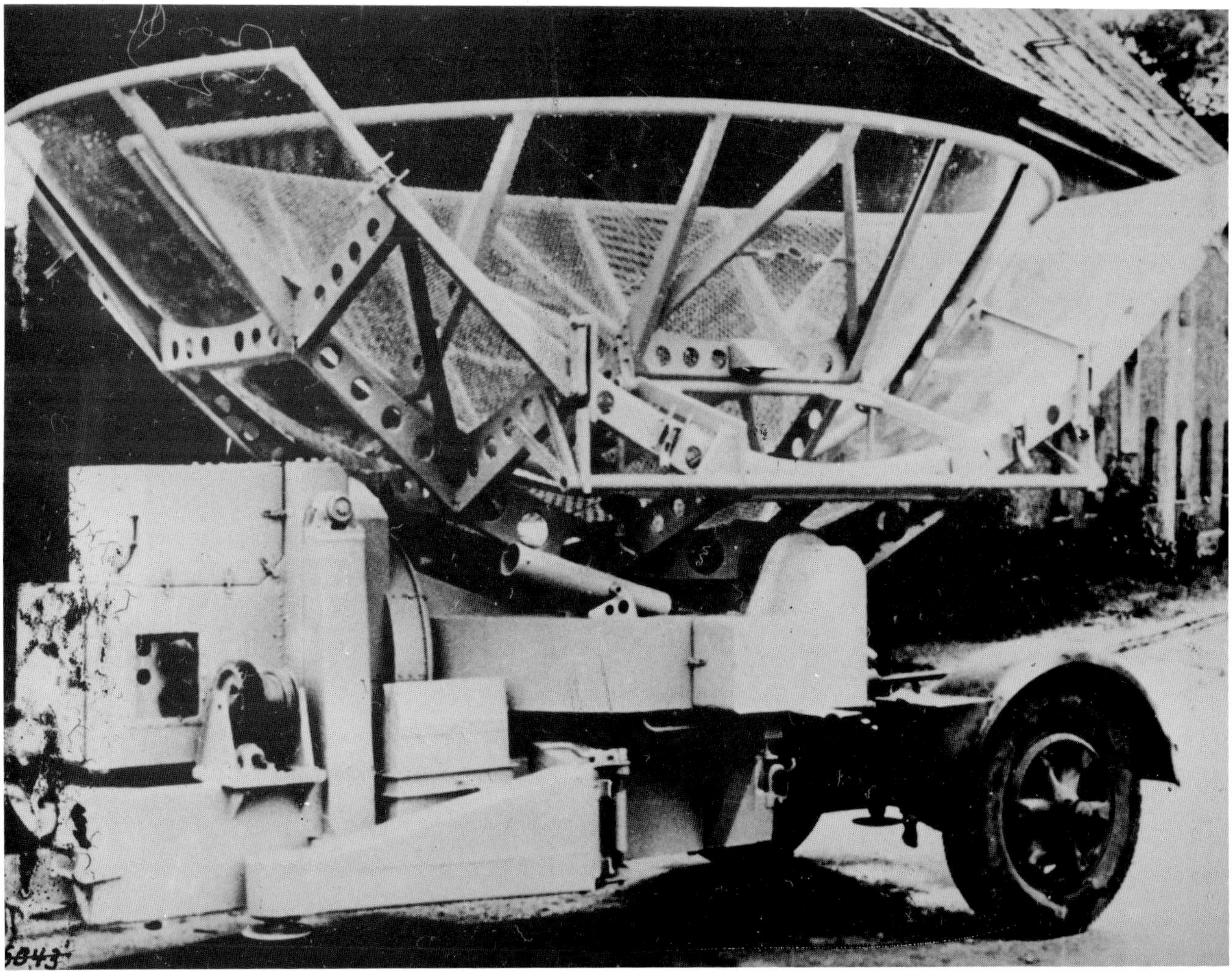

night of 24/25 July 1943 the city of Hamburg was the victim of a large scale bombing raid, during which over 40 tons of aluminum foil were dropped. Both the Flak's Würzburg radars as well as the Lichtenstein systems of the night fighters were put out of operation because of the number of blips on their screens. The crews were not able to distinguish aircraft echoes from the chaff returns appearing on their scopes, thus preventing a concentrated AAA fire or effective night fighter operations. The Freya fire control/aircraft warning systems operated on 125 MHz (corresponding to a 2.4 wavelength), meaning that the chaff had to be 1.2 meters long. The chaff would often become lumped together and rapidly fall to the earth. Until this problem was solved the Freya radars were jammed by transmitters on board special aircraft within the bomber streams. The results of the unhindered bombing raid was devastating for Hamburg, while according to their own accounts the British had never suffered so few losses as they did on this mission.

Those anti-jamming measures which were then developed can only be mentioned briefly here, and no attempt is made to describe their precise functions.

One of the first measures against chaff interference undertaken by the Germans was the installation of the Würzlaus set in the FuMG 62. It operated by making visible the radio frequency of the Doppler effect, i.e. the DF echo from fast moving targets, which appeared as bright moving blips on the radar scope, stood out against the chaff drifting slowly downward. However, operating a radar system fitted with the Würzlaus was difficult in that prior to each operation the radar had to be calibrated against a calibration point. Furthermore, it was only effective against a relatively small number of tinfoil strips. With the Tastlaus set or Taunus switching the chaff blips appeared fainter on the radar scope than the returns from aircraft. The Nürnberg system utilized a listening set to feed the Doppler modulations from the propeller noises acoustically to the headphones of the ranging crewman at the radar. The brightness and clarity of the blip on the scope had to be optimal in order to achieve an accurate plot. From 1944 on, the most effective way to eliminate chaff interference on the FuMG 62 without affecting ranging performance was through the use of the K-Laus in conjunction with the Windlaus. This anti-jamming system could also be used with the Würzburg-Riese radar and, with a few modifications, for the FuMG 64 Mannheim.

Direct jamming of the transmission frequencies on German systems could be avoided by rapidly switching the frequencies. Accordingly, a modification program under the codename "Wismar" was undertaken which saw the narrow band antennas replaced by wideband dipoles, the Eidechse transmitter component by the Urechse and the receiver oscillator by the heterodyne oscillator known as Michael.

Yet the "radar war" continued. Up to this point the flak and aircraft warning systems which operated at greater wavelengths than the Würzburg radars had not been affected by chaff jamming. In May of 1944 the enemy began dropping 1.20 m long tinfoil strips in large quantities, which had the effect of jamming not only the Freya systems, but also the Wassermann and Mammut radars as well. The Jagdschloß air surveillance radars were also affected. Within a short time, however, the Freya and Wassermann radars were fitted with an anti-jamming device, the Freya-Laus and the Wasserfloh, respectively. This permitted the chaff echoes to be distinguished from the aircraft blips on the screen. However, an anti-jamming system for the Jagdschloß radar had not been put into production by the time the war ended.

The most effective countermeasure against jammers was the introduction of systems which operated in the superhigh frequency area of 3,000 - 30,000 MHz and the 9 cm wavelength, shortly supplemented by the 3 cm wavelength.

Although successful experiments had been carried out earlier in this region, higher authorities prohibited further development along these lines. It was only due to the second significant event that attention once again became focused in this area. A system which operated in the superhigh frequencies was found on a British bomber which had been shot down over Rotterdam. It was discovered that with the aid of this H2S air-to-ground panorama system ground targets could be distinguished with great accuracy, even through cloud cover.

A "virtual map" of the region being overflown was projected onto the screen, which the so-called pathfinder aircraft used to navigate by and drop flares over the target. The following bombers then would unload their bombs over the flare markers.

Based on the Rotterdam system find, named by the Germans after the area in which it was recovered, scientists now began work anew on experiments with systems operating in the superhigh frequency range, first for a bomber navigation system, then also for the Flak arm as a search and track system. These experiments now had the approval from the highest authorities, not in the least because it was revealed that the Rotterdam system was virtually impervious to jamming, even when exposed to massive amounts of chaff.

FuMG 77 Marbach V

The first truly functional German radar which operated in the superhigh frequency region was the FuMG 77 Marbach V "Rotterheim". It consisted of a Mannheim revolving stand and reflector, the instrumentation from the Würzburg system, and the new Rotterdam transmitter and receiver, consisting of a rotating 9 cm dipole. Having a ranging accuracy of ±25-35 m, an azimuth accuracy of ±1-2/- and a vertical accuracy of ±1-2/16°, the system was suited quite well as a fire control radar for the Flak. The range against a target flying at 3,000 meters' altitude was around 27 -30 km, depending on aircraft type. Bomber raids at 10,000 meters could be tracked out to 35 km. Due to the concentrated focus of the beam (a halfpower width of around ±2°) the Einweisungsempfänger 37 data receiver was needed to acquire targets.

Here we see the FuMG 76 Marbach with its 4.5 meter dish and the "centimeter grill". The FuMG 77 Marbach V "Rottenheim" kept the control stand from the FuMG 64, while its electrical ranging component was removed and the FuMG 62 D's instrumentation was used in its place. This radar type was only produced in a small number of hand-built systems.

The Egerland System with FuMG 74 Kulmbach and FuMG 76 Marbach

The most capable superhigh frequency AAA fire control system at the time was the Egerland radar combination, developed over the course of 1944. It consisted of the FuMG 74 Kulmbach air surveillance radar, the FuMG 76 Marbach tracking radar and the Bedienungswagen 74/76 Bayern operations van with the system sets, which could be either separately entrenched or housed in a bunker.

With the exception of the antenna the FuMG 74 Kulmbach was adopted with few changes from the Berlin A aircraft radar. As an omnidirectional air surveillance radar it painted a panoramic picture by sweeping the airspace out to around 50 kilometers, with aircraft being acquired from 20 to 30 km out. It operated in the 9 cm wavelength and was accordingly fitted with a relatively short 6 m tube antenna. This was in the shape of a narrow vertical fan antenna which rotated around the vertical radar shaft at 20 rpms.

Targets which had been acquired were passed off to the FuMG 76 Marbach for precise location plotting. This radar also operated in the 9 cm wavelength and, by utilizing the 4.5 meter reflector from the FuMG 68 Ansbach, achieved quite a high directional accuracy. The range was around 30 - 50 km, ranging accuracy was ±25-35 m, azimuth accuracy was ±0.7-1/- and vertical accuracy was ±0.7-1/16°.

The central controls for both radars, plus their instrumentation equipment, was located inside the separate Bedienungswagen 74/76 Bayern operations van.

If a target appearing on the omnidirectional scope of the FuMG 74 needed to be pinpointed with greater accuracy, it was marked with a revolving pointer which was linked to the transmitter from the Übertragungsgerät 37 and from there the target's position was fed to the Empfänger 37 receiver on the FuMG 76. With the operator covering both pointers the FuMG 76 was aimed in the predetermined direction of the target with the aid of the AEG-Leonard control. Cooperation between the two radar systems went so smoothly that it was possible to pass off up to seven tracks per minute. During daylight, the azimuth and elevation of the FuMG 76 could be controlled with the assistance of an optical range-finding telescope. The Übertragungsgerät 37 was also used to feed the precise target location data from the Marbach system to the Kdo.Ger.40 of a AAA battery.

Although only two prototypes were operationally deployed, this Egerland system proved to be the optimal radar system for the Flak arm.

The Egerland system was a Flak surveillance and firing system. It constituted the FuMG 74 Kulmbach air surveillance radar, on the right, the FuMG 76 Marbach fire control radar on the left, and the Bayern operations van for remotely operating the radars. This was buried between the two radars and cannot be seen in this photograph. This site here was located in Teltow near Berlin

FuMG 75 Mannheim-Riese

The FuMG 74/76's accuracy and security against chaff jamming was only attained by the FuMG 75 Mannheim-Riese, although the latter was only built in small numbers. This system comprised the measuring components from the FuMG 64 Mannheim and the revolving base from the FuMG 65 Würzburg-Riese. The resulting FuMG 75 Mannheim-Riese had a supplemental control stand attached to the right side, which held the sighting optics. A target could be precisely followed using this optical sighting system. The radar dish, coupled to the sight, was simultaneously directed to the target. The range was increased to 84 km, distance accuracy was ±12 m and the azimuth/vertical ranging accuracy was ±1-1.5/- and ±0.7-1/16°. This system was planned for guiding anti-aircraft missiles, although the development of these missiles did not reach operational status before the war's end.

Above: The controls for both radars and the instrumentation were housed in the Bedienungswagen 74/76 Bayern operations van.

Right: The FuMG 74 Kulmbach operated as an omnidirectional air surveillance radar. By constantly scanning the airspace it provided a panoramic picture of the environment and all aircraft contained therein. It operated on the 9 cm wavelength and could pick up targets out to a distance of around 50 km.

Right: Here we see the complete Egerland system with the FuMG 76 Marbach targeting radar (left), the FuMG 74 Kulmbach air surveillance radar (right) and the entrenched Bayern 74/76 operations van in the center.

Air Warning Radar Systems

Early recognition of approaching enemy aircraft was important to the Flak, the fighter pilots, the Marine and naturally also for civil air defense. At the beginning of the war there were only the previously mentioned Freya systems available, having a range out to about 80 km and, due to a shortage of other types, these were also used for directing flak batteries and for fighter direction. These systems, not all of which are included here, were constantly being seeing improvements in their range and accuracy.

In addition, variants of the FuMG 43 "Freya-Fahrstuhl" (Freya Elevator) saw operational service. These systems prevented high-flying approaching targets from disappearing in the ground reflection nulls of the vertical plane directional pattern while at the same time providing a rough estimation of the altitude.

When the first anti-Freya jammers began appearing in 1942, countermeasures immediately followed, which in the main consisted of switching transmitters and receivers to different frequencies. The Freya-Laus was in production by 1943 to counter chaff jamming.

The companies of AEG and Telefunken produced the next series with a spot frequency in the area of 162 - 200 MHz. For the Luftwaffe, these were given the designation of FuMG 451 Freiburg II and for the Marine FuMG 321 - 328. These provided target information from 20 km out to 130 km.

For the higher wavelengths Yagi antennas were used, bearing a strong resemblance to the television antennas of today. A total of 14 different "Köthen frequencies" ranging between 34.5 and 177 MHz are known to have existed for these Yagi-Freya systems.

In 1941 the first experiments were carried out with two Freya radar systems in directing fighters using the so-called "Erstling" method. This involved one radar tracking the enemy, while the other plotted the German fighter. Updated locations were entered onto a map and intercept courses to the enemy were passed by voice radio to the German fighter. This system was replaced by the so-called EGON system in 1943. (EGON = Erstling-Gemse-Offensive-Navigation method). Since this method saw several improvements and updates, we will not discuss it in further detail.

This aerial photograph shows the Freya early warning system at Cap Blanc-Nez (134 m) with the memorial to the fallen soldiers of the First World War, approximately 20 km southwest of Calais.

Above left and right: Among the numerous Freya radar variants are these two examples of Freya systems with Yagi antennas, which were planned to operate at higher wavelengths. Of these so-called Yagi-Freya systems, 14 different radars are known to have existed with "Köthen frequencies" between 34.5 and 177 MHz (Köthen frequencies were named after the AAA regiment stationed in the town of Köthen which was entrusted with conversion to the Freya radars.)

The radar on the right in this photo is a special variant of the Freya LZ system for EGON fighter direction. The suffix LZ indicates that the system could be disassembled for airborne transport and EGON stood for Erstling-Gemse-Offensive-Navigation method, a system for controlling bombers and ground attack aircraft using a ground-based system. It consisted of a cross base, a revolving shaft with the mounts for the system, a shack which could be dismantled for transport and a removable antenna. A Würzburg-Riese is on the left.

Above left: The fighter direction post on the island of Wangerooge utilized experimental equipment for improving its height finding capabilities. The "Freya Fahrstuhl", as this radar type was called, consisted of the receiver antenna from a Freya radar, which could be raised and lowered at will up and down its two masts. In so doing, it prevented the disappearance of approaching targets flying at high altitude as they passed over the ground reflection nulls of the vertical plane directional pattern.

Above right and left: With the second version of the "Freya Fahrstuhl" the new horizontally polarized Freya antenna could be raised and lowered on a 6-meter wide and 20-meter high frame. These radars could also provide an approximation of a target's altitude. The range was dependent on the aircraft's altitude and the elevation of the radar site. With an altitude of 2000 meters the range was 65 km; at 8000 meters a target could be picked up as far away as 185 km.

Above left and right: Here we see the third style of the "Freya Fahrstuhl", by which the antenna could move vertically along the rail of a 20-meter high trellis mast.

Right: The crew of this "Freya Fahrstuhl" were female Luftwaffe crew members, or Luftwaffehelferinnen, who are posing here for a scrap-book photo on the roof of the operations van.

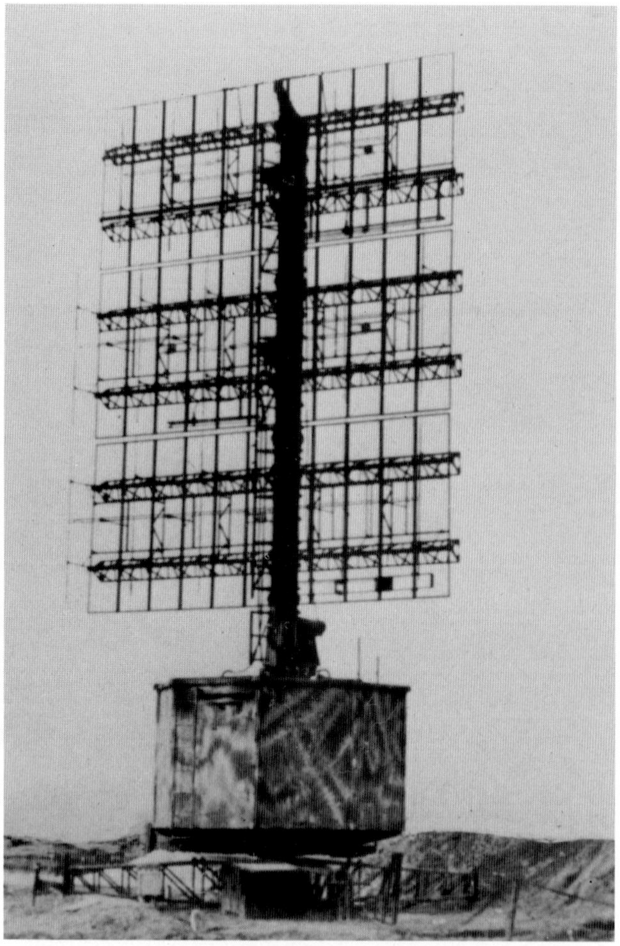

Above left: This FuMG 401 A Freya LZ has its IFF antenna on the backside of the antenna grid. The radar operated in the frequency range of 120 - 158 MHz and had a range of 30 to 150 km.

Above right: The FuMG (Flum) 41 Freiburg I with "Radattel" DF was used by the Marine as an air warning radar. The Radattel DF system was a special DF system introduced by the GEMA company in 1942 and got its name from the sound the motor which was used to power the system made when operating!

Left: The significance of the upper antenna on this Freya radar is no longer known. This was probably a prototype radar. (Or can someone shed additional light on its application?)

Long-Range Radar Prototypes

Wassermann L and S Systems
FuMG (Flum) 41,42

Just after the outbreak of the war, a requirement was issued for a radar system for early warning whose range would be significantly greater than that of the Freya. In order to accommodate this requirement, four Freya antennas were initially set up one on top of the other on a 36-meter high trellis mast. This in turn was mounted to a rotating stand. Tension wires ran from a pivot bearing at the top of the mast to the ground to provide stabilization. In the center of the antenna field was an area for the IFF system. The radar was controlled by the Leonard remote control system. This "Wassermann L" (L = leicht, or light) operated next to a Freya radar (125 MHz, 20 kW) and was supplemented by a phase shifter with an elevation scope. The actual range was 200 km and the azimuth accuracy was ±4°.

The Wassermann S (S = schwer, or heavy) had double the number of antenna fields, attached to a self-supporting steel tube rotating tower having a diameter of 4 meters and a height of 60 meters. This stood in a concrete base and could rotate with the aid of a large toothed gear drive. Even by using more powerful transmitters of 100 kW and greater the range of the Wassermann S radar, at around 300 km, was less than hoped for. The azimuth accuracy was ±3°.

FuMG 402 I to V
Wassermann M I to M V

The firm of Siemens produced improved variants of the Wassermann radars beginning in 1942. Its first Wassermann M I, FuMG 402, had a 36-meter high trellis mast and was similar in design to the L version. However, it displayed the A/N azimuth DF and a few other improvements.

The transmitter mast for the Wassermann M II radar, FuMG 402 II, was 40 meters high and had a wider, horizontally polarized antenna plus the "Kuh/Gemse" IFF system and its associated antenna. Both systems stood in a concrete foundation which allowed them to pivot.

With the Wassermann M IV and M V, FuMG 402 IV and V, the operations van was located in the center of the former's 51-meter and the latter's 60-meter high masts. The radar operated in the frequency range of 120 to 158 MHz with an output of 100 kW. Its range was 300 km, although over water a range of 380 km could be achieved. Azimuth accuracy was ±25°; vertical accuracy was ±75°.

The FuMG 402 Wassermann M, M III and M IV systems, had a height of 51 meters. As a long-range search radar it was not only used in the air warning role, but also found application with the EGON system for fighter direction control. Its maximum range was around 350 km.

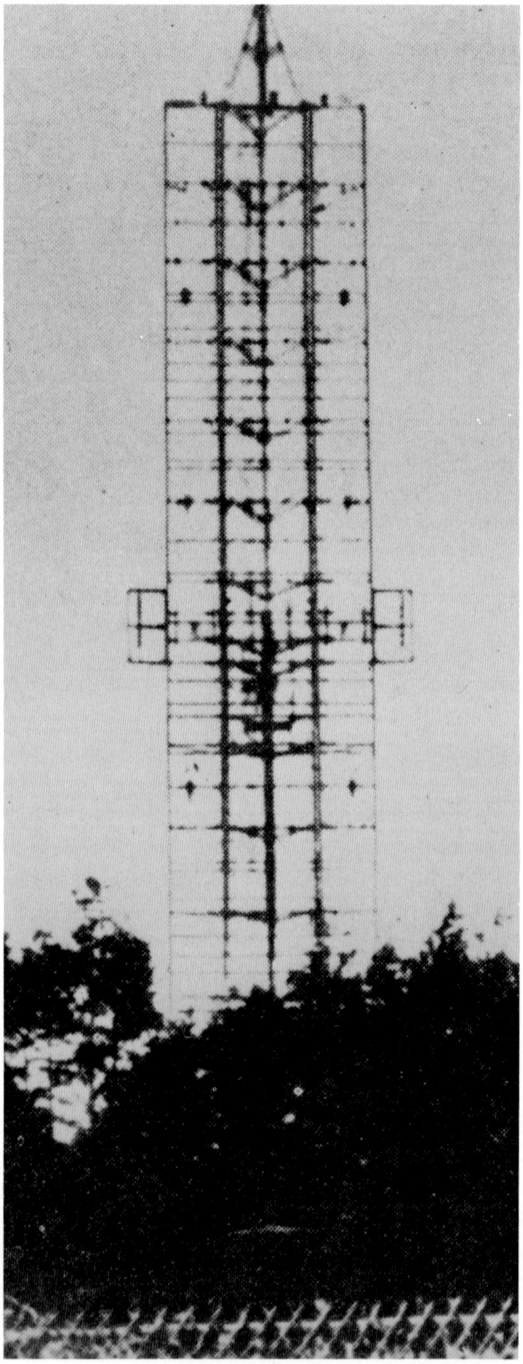

Above left: On the FuMG 402/IV Wassermann M IV the operations van was no longer located at the base of the 51-meter antenna mast, but instead was moved to the center of the antenna.

Above right: The four Freya antennas of the FuMG 41 Wassermann L were arranged vertically and attached to a trellis mast 36 meters high. The system worked with the standard Freya radar set plus an additional phase shifter, or compensator. A somewhat wider antenna can be seen in the center of the array, the purpose of which was for IFF signals reception.

Left: Here the pivot base of the Wassermann M IV can be seen to good effect on the lower part of the radar.

Above: This Würzmann experimental radar was set up near Göhren on the island of Rügen. Its antenna was mounted to two 36-meter high trellis masts and consisted of two stacked antenna arrays, each having 16 panels from the "Michael" medium wave directive beacon radio system. Against aircraft the system had a range of around 200 km. This radar was used to pick up the first echoes from the moon in 1944, which required around 2.5 seconds for the 384,000 km journey. Next to the house on the right is a Kurpfalz or Kurmark radar.

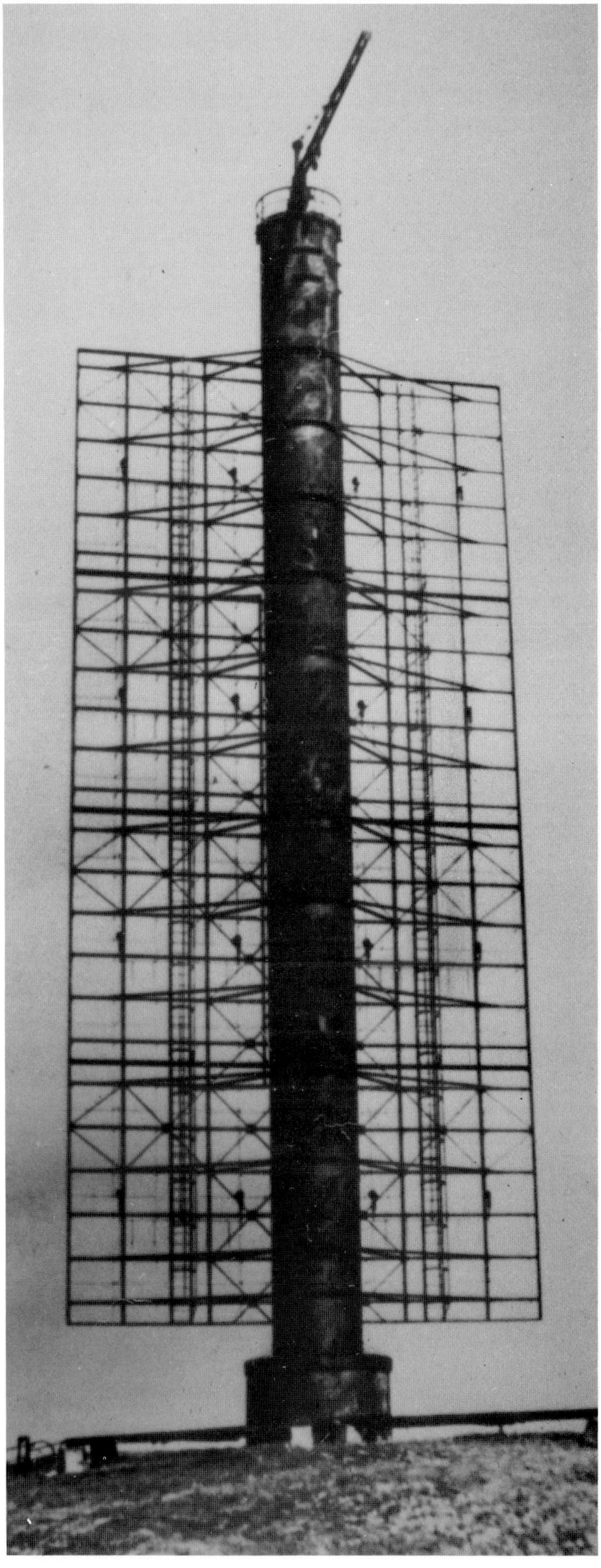

Right: Although the FuMG 42 Wassermann S was fitted with the same radar set as the Wassermann L, it had double the amount of antenna arrays as the L-series mounted to its 4-meter diameter and 60-meter high rotating steel shaft. The radar seen in this photograph lacks the uppermost antenna array, which served as the IFF receiver.

Above: The Luftwaffe's Mammut radars had four trellis masts, while those of the Kriegsmarine had only three. Here we see a Mammut II-Friedrich Luftwaffe radar, which had an antenna 11.1 meters tall and 28.5 meters wide. It operated with 100 kW transmitters. Some of these radars sited along the main approach and exit corridors were fitted with two antenna systems back to back, as in this photo. The radar was located on bunkers in which the crew sat operating the equipment. These radars were primarily set up in the coastal regions.

Mammut I and Mammut II Long-Range Search Radars

The GEMA company also became involved in the development of early warning type radars. Its FuMG (Flum) 41 G Mammut I had four trellis masts in a row. Attached to these were eight Freya antennas which together made up a field 10 meters tall and 25 meters wide; the upper section acted as the transmitter and the lower section as the receiver. The follow-on Mammut II radars also had four trellis masts, but the mounted antennas covered an area 11.1 meters by 28.5 meters with 100 kW transmitters. In order for the radar to pick up the primary approach and departure courses, some of these systems were fitted with an antenna array both on their front and on their back. The radar was able to pick up targets at 8,000 meters out to a range of 300 km. Low-flying targets at around 50 meters' height, however, were acquired at around 35 km. Range accuracy was ±300 m and the azimuth accuracy was ±0.5°.

Left: With the Mammut early warning radar the GEMA company created a 10-meter high and 25-meter wide antenna field attached to four trellis masts. This antenna was built from Freya antennas. The photo shows one of the first prototypes prior to being wired up.

Above: This Mammut early warning radar was located near Lorent and is seen here still under construction.

Beyond-the-Horizon Experimental Radars

Only a few of the more interesting beyond-the-horizon experimental systems will be briefly discussed here in order to provide a glimpse into the skills of the radar technicians.

Since ultrasonic waves could only be transmitted along a line of sight, the horizon acted as somewhat of a barrier for distance measuring. Therefore, beyond-the-horizon systems made use of short wave signals. One of the first experimental systems developed was the Knickebein J (J = Jonosphäre, or ionosphere), which was a guide beam installation built in 1941. This system had a range of 0 - 300 km with a PRF of 500 Hz. But after switching the impulsing to 50 Hz/25 Hz it had a range of 0 - 6,000 km, i.e. echoes were picked up out to a distance of 3,000 km, although the quality varied greatly depending on the season and time of day.

Rundblick was an early warning radar with a 360° field of coverage, operating at 26.2 MHz. A 40-meter wide antenna was attached in front of a rotating shack, whose center section contained the receiver. Ships and low-flying targets could be picked up out to 60 km with this system. High-flying bomber formations could be tracked as far out as 230 km.

Below: The Rundblick early warning radar was mounted in front of a rotating cabin. Its antenna consisted of a 40-meter wide reflector "wall" with eight vertical dipoles for duplex operations, i.e. the antenna's surface was used by both the transmitter and receiver.

The Elefant and See-Elefant systems only saw limited service. The transmitter for the Elefant was a 90-meter high mast with a broadcast-type antenna. Reception and the maximum triangulation were accomplished by means of cross arms supporting six horizontal line arrays on a rotating Wassermann M IV mast. The range was around 200 - 300 km. The "Max" station on the Dutch coast was known to have located a convoy sailing near the island of Jan Mayen in 1944 while operating in beyond-the-horizon mode, a distance of some 2,200 km.

Left: This transmission tower was part of a limited number of Elefant beyond-the-horizon radar systems, here seen at the "Max" site near Castricum in Holland. Its transmitting broadcast-type antenna was fitted to a 90-meter high, fixed wooden mast.

Below: The receiving tower for the Elefant system consisted of a 51-meter high Wassermann M IV steel mast, to which were fitted extension arms holding six horizontal line arrays. On the left we see the front view and on the right a rear three-quarter perspective. The cabin for the crew and equipment was located in the center of the mast.

The See-Elefant early warning radar at the "Robbe" station on the island of Röm in Denmark was a very versatile system. The wire-braced transmission antenna was located between two fixed 100-meter high masts, while the transmitter itself was in an independent Mammut system at the foot of the masts. Two DF receivers were located one kilometer on either side of the transmitter. Each of these 70-meter high Wassermann masts had eight horizontal line arrays fitted one above the other. The system's range was 250 - 400 km. However, when operating in beyond-the-horizon mode signals could be picked up out to 4,000 km. Range accuracy was said to have been ±1 to 2 km, the azimuth accuracy at ±5°, later increasing to ±1°.

This system was used to measure the final segments of the V2 missiles' flight paths over London and register their impacts.

These three pictures show the Robbe experimental site. It was located with its See-Elefant radar on the Danish island of Röm. The fixed transmitting antenna was wire-braced on two 100-meter high masts. The transmitter was housed separately in the systems bunker of the Mammut radar located below (photo above right). At about 1000 meters on either side of this radar were the two DF receivers on 70-meter high Wassermann masts having eight line arrays arranged vertically (below photos).

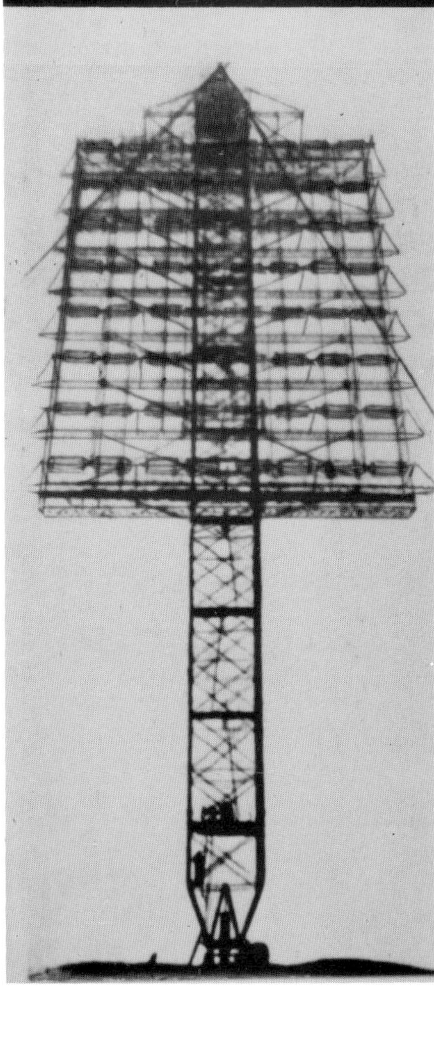

The Klein-Heidelberg (Little Heidelberg) radar is also worth mentioning when discussing early warning systems. For reasons of camouflage it was fitted to the back of a Wassermann S antenna. It was given a supplementary name of "Parasit" (parasite) because it could also be used to detect those aircraft which the British were tracking. This was possible because the transmitting waves which the British were sending were known and these were not only reflected back in the direction of the transmitter. The detection range was around 250 km, range accuracy was ±1 to 2 km and the azimuth accuracy was ±3°, later ±1°. The system functioned smoothly even when the enemy used jamming signals and chaff countermeasures.

Below: The DF antenna for the Klein-Heidelberg early warning antenna was fitted to a Wassermann S tower. The original style is seen here on the back of the mast. This system included a synchronizer antenna with receiver, which stood on a simple wooden mast nearby. One of the biggest advantages of this system was the fact that it could provide location data even under the most severe jamming conditions.

Below: Here is another photo showing the lower section of the receiver tower for an Elefant system, with the operations cabin in the mast's center.

Omnidirectional Systems

The radar systems which we've discussed up to this point could only "see" targets that moved within a relatively narrow sector in front of the radar. Because of this, the need for an "omnidirectional" or "panorama" radar was soon felt which could acquire all targets in a 360° area around the radar. An example of this is the FuMG 404 Jagdschloß, proposed by Siemens in 1943. 44 of these systems were in operation by the time the war ended. A rotating wideband antenna with 4 x 16 horizontal dipoles revolved at a height of 7 meters above a fixed hut. Above this antenna was an additional wideband antenna for IFF interrogation and signal reception. This fixed site omnidirectional radar had a range of 80 to 200 km with an output of 150 kW. Other omnidirectional radars such as the Jagdschloß X, Forsthaus, Jagdwagen, and the Jagdhütte were only produced as prototypes and never achieved operational status.

The antenna of the FuMG 404 Jagdschloß was installed on a fixed hut. On top of this rotating wideband antenna, with dimensions of 3 x 24 meters, rested a second smaller wideband antenna. The latter was vertically polarized and was used for IFF interrogation and signal reception.

These two pictures show the Bernhard omnidirectional radio ranging system made by Telefunken, which was used as a navigation aid for bombers and night fighters. It was therefore not a true radio detection or radio location system, but was instead a beacon transmitter site. A detailed description of the system's operation is not within the scope of this book.